General Rules of Marine Insurance 1919

adopted by the German Underwriters
drafted in collaboration with German Chambers of Commerce
and other Corporations concerned
under the auspices of the Hamburg Chamber of Commerce.

Authorised Translation
prepared by
DR. ALFRED SIEVEKING, lawyer in Hamburg
at the request of and published by
The Corporation of Hamburg Underwriters.

HAMBURG
L. Friederichsen & Co.
1920.

Printed by LÜTCKE & WULFF, Hamburg.

CONTENTS.

First Part. General Principles.

Second Part. **Special Rules on special Subjects.**

Section 1. Insurance on Ship.

Section 2. Insurance on Goods.

FIRST PART.
General Principles.

SECTION 1.
Insurable Interest.

§ 1.
Insurable interest defined.

(1) Every person has an insurable interest, who stands in any relation to a marine adventure in consequence of which he may benefit by the safety or due arrival of ship or goods at risk therein, provided that the loss or damage sustained can be measured in money.

(2) The principal subject-matters of marine insurance are as follows:

- the ship,
- goods,
- profit contingent upon the arrival of goods at place of destination (prospective profits),
- commission to be earned in case of arrival of ship or goods at place of destination,
- freight,
- ship's hire,
- passage money,
- loan on bottomry and respondentia,
- advances, disbursements and other liabilities, if the amount of the debt is secured by property exposed to marine risks,
- the risk undertaken by the insurer (reinsurance).

A contract expressing to be on one of the aforementioned subject-matters of insurance does not cover a subject-matter not designated therein. A reinsurance, in particular, must be specified as such at the time the contract was entered into.

(3) A wrong statement as to the interest in the thing insured entitles the insurer to avoid the contract. A designation, however, of ship's hire as time-freight or vice versa does not entitle the insurer to avoid the contract.

§ 2.
Where the assured has no insurable interest.

(1) A contract is void, where the assured has no insurable interest. Thus, a contract by way of gaming or wagering is void.

(2) Nevertheless, the premium is due, unless the insurer was aware of this fact at the time the contract was entered into.

§ 3.

Insurable interest and payment of premium.

(1) The person effecting the insurance*) is discharged from his liability to pay the premium, if, at the time the contract was entered into, he did not know or need not have been aware of the fact that the assured had no insurable interest. If the contract has been concluded through an agent, the agent to insure is deemed to know whatever, in this respect, was known to or ought to have been known by the person effecting the insurance.
(2) If, however, the person effecting the insurance does not disclose the fact that there is no insurable interest within a year from the day the contract has been concluded, or if, having become aware of this fact within the course of the year, he does not immediately disclose it to the insurer, he remains bound to pay the premium.
(3) If no premium is due, the insurer, nevertheless, is entitled to receive a consideration to the amount of one half of the premium, but in no case exceeding $^1/_8$ per cent. of the sum insured (vide infra § 18).

§ 4.

Where interest ceases or does not come into existence.

(1) The person effecting the insurance is discharged from his liability to pay the premium, where the interest insured has not attached or where the assured, though in treaty for, has not acquired an interest in the thing insured, provided these facts have been disclosed to the insurer by the person effecting the insurance within a year from the day the contract was entered into or immediately after. If no premium is due, the insurer is entitled to receive a consideration within the limits defined in § 3, subs. 3.
(2) Where the interest has attached, the premium is due, although the interest subsequently may have ceased.

---

*) The person effecting the insurance (Versicherungsnehmer, assuré contractant or commissionnaire) is the person, in whose name the contract of insurance is made. If this person covers his own interest in the subject-matter insured, the person effecting the insurance is identical with the assured. If he covers the interest of other persons (insurance for account of others), not acting in his own name, he is simply an agent and is not termed a person effecting the insurance. A broker, generally, is an agent; he is, however, a person effecting the insurance, if the contract is made out in his name. For particulars vide infra §§ 52 fg.

§ 5.

Lost or not lost.

(1) The parties may agree that the risk shall attach at a date prior to the agreement. No premium, however, is due where the insurer at the time when he entered into the agreement knew that the subject-matter insured was no longer exposed to the risks insured against. If, on the contrary, the person effecting the insurance, at the time when he entered into the agreement, knew or ought to have known that the event insured against had already occurred, the insurer is discharged from liability; the premium, however, is due, where the insurer was not aware of the loss at the time the agreement was made.
(2) Where the contract has been concluded through an agent, the agent to insure is deemed to know whatever was known to or ought to have been known by his principal.

SECTION 2.

Insurable Value. Undervaluation. Overvaluation. Double-Insurance.

§ 6.

Insurable value.

(1) The insurable value is the full value of the subject-matter insured.
(2) Where the parties agree as to the value to be placed upon the subject-matter insured, the valuation is conclusive of the value of the interest insured. In a case of excessive valuation the insurer is entitled to have the valuation corrected. Where the assured has been insured for an amount less than the valuation agreed upon, the insurer is liable for such proportion of the measure of indemnity as the sum insured bears to the full extent of the value originally agreed upon; this rule applies also to the case where the insurer has successfully contested the valuation.
(3) A valuation only provisionally agreed upon does not constitute a valuation within the meaning of these rules.

§ 7.

Separate valuation.

(1) Where different species of property have been insured in a single sum with a distinct valuation on each, the person effecting the insurance is entitled to claim for each species as being separately insured. This rule, in particular, applies to the case where sub-divisions of goods of an homogeneous kind have been separately valued (series).

(2) The rule of subs. 1 does not apply to the case where the different series the goods belong to cannot be ascertained.
(3) Likewise the rule does not apply, where the series shall be formed according to the order in which the goods leave the ship, when properly discharged, in a case where the order, in which they have been discharged, has not been ascertained at or immediately after the discharge by running landing numbers or in a similar way. The series are to be so formed, only if this has been specially agreed upon.

§ 8.
Under-insurance.

Where an insurance has been taken for an amount less than the insurable value, the person effecting the insurance is deemed to be his own insurer in respect of the uninsured balance. The insurer, in such a case, especially as regards loss or damages or expenses incurred, is liable for such proportion only of the measure of indemnity as the sum insured bears to the insurable value.

§ 9.
Over-insurance.

(1) Where the insured value exceeds the value of the thing insured, the insurance is void to the extent of the over-insurance. As to a return of premium and the consideration due to the insurer, the rules of § 2, subs. 2, and § 3 apply.
(2) In case of fraud on the part of the person effecting the insurance, the contract is void. The insurer, however, is entitled to the full amount of premium, unless he himself was cognizant of the fact at the time the contract was made.

§ 10.
Double-insurance.

(1) Where several contracts of insurance have been effected on the same adventure and interest, and the sums insured exceed the insurable value, the insurers are conjointly liable, that is to say: that the person effecting the insurance is entitled to claim from each insurer the whole amount he has subscribed, provided he does not claim a sum exceeding the amount of the loss or damage sustained.
(2) As between themselves, the loss is apportioned according to the sums each insurer is liable to pay upon the contract he made with the person effecting the insurance. This rule is applicable also in a case, where one of the contracts of insurance is governed

by Foreign law, provided a like rateable distribution is prescribed by the rules of this law.
(3) In case of fraud on the part of the person effecting the insurance, the contract fraudulently entered into is void. The insurer, however, is entitled to the full amount of premium, unless he himself was cognizant of the fact at the time the contract was made.

§ 11.

When double-insurance may be rectified.

(1) Where, unaware of a contract already existing, a person effects an insurance thereby causing the risk to be doubly insured, the insurers, provided the risk has not yet attached, are bound to submit to the sums respectively insured being reduced in proportion to the amounts each of them is liable to pay upon his own contract; in such a case, also a proportionate part only of the premium is due. The person effecting the insurance, however, is debarred from claiming such a reduction, if he does not assert his claim immediately after he has become cognizant of the fact of a double-insurance having been concluded.
(2) Where the insurer is obliged to submit to the premium being reduced, he is entitled to receive a consideration within the limits defined in § 3, subs. 3.

§ 12.

Notice to be given in case of double-insurance.

The person effecting the insurance having become aware of the fact that a double insurance has been concluded, is bound immediately to give notice thereof to the insurer.

SECTION 3.
Insurance is uberrimae fidei.

§ 13.

A contract of marine insurance is a contract based upon the utmost good faith, to be observed by all parties concerned.

SECTION 4.
The Policy. The Premium. Return of Premium.

§ 14.

The policy.

(1) The insurer, if asked to do so, shall hand to the person effecting the insurance a document signed by himself and embodying the contract of marine insurance. This document is called The Policy.
(2) Where a policy has been issued, the insurer is only bound to

pay upon the document being produced. He is discharged by paying to the bearer.

3) Where the policy has been lost or destroyed, the insurer may refuse payment, until the policy has been invalidated by Decree of Court or securities, other than guaranties, have been tendered. The same conditions must be complied with, before the insurer can be asked to issue a copy of the policy; the costs of issueing a copy are to be born by the person effecting the insurance.

§ 15.

Terms of policy.

The person effecting the insurance is deemed to have approved of the terms of the policy, unless he denies having done so immediately after having received the document. Nevertheless, it remains open to him to repudiate his approval upon the plea of mistake in accordance with the general rules of the law (Civil Code §§ 119 fg.).

§ 16.

Premium when payable. Securities when required.

(1) The premium, including additional charges, is due upon the conclusion of the contract. This is true also in the case where, according to the general custom of trade or the custom established in the business relations between the insurer and the person effecting the insurance, payment usually is made at a later date.

(2) Notwithstanding the fact that a certain date has been agreed upon or respite has been given for the payment of the premium, the premium becomes payable immediately upon the termination of the risk. The person effecting the insurance, however, in allowed to make a set off with all claims for compensation, whether due or not, he has against the insurer and which accrue from the same relations out of which arises the claim of the insurer on account of the premium.

(3) Where a certain date has been agreed upon or respite has been given for the payment of the premium, the person effecting the insurance is bound, if called upon, to tender securities in a case where the chance of his being able to pay the premium has been impaired through a material change in his circumstances.

§ 17.

Failure in paying the premium or in tendering securities.

Where the premium has become due or securities have to be tendered and the person effecting the insurance does not comply with a call of the insurer to pay the premium or to tender the securities within a certain reasonable time, the insurer is discharged from liability, if loss or damage occur before the premium thus called for has been paid or securities have been tendered. In such a case, the insurer is also entitled, if the premium has not been paid or the securities have not been tendered within the prescribed time, to cancel the agreement without warning and, upon a voyage policy, to claim the whole amount of the premium or, in case a time policy has been effected, the consideration as defined in § 3, subs. 3.

§ 18.

Consideration for return of premium.

In cases of return of premium the consideration due to the insurer amounts to a sum varying between one half of the premium and a sum not exceeding $^1/_8$ per cent. of the sum insured.

SECTION 5.

Disclosure. Representations. Alteration of Risk.

§ 19.

Disclosure.

(1) The person effecting the insurance must disclose to the insurer, before the contract is concluded, every material circumstance which is known to him, including all informations with regard to material circumstances received by him, whether he holds them to be true or trustworthy or not. Every circumstance is material which would influence the judgment of a prudent insurer in fixing the premium or determining whether or not he will take the risk. Matters of common knowledge need not be disclosed.
(2) The person effecting the insurance must transmit to the insurer, before the contract is concluded, every circumstance, which must be disclosed, as speedily as is consistent with the ordinary course of business. As a general rule, such a disclosure must be transmitted in the same or in a similar mode to which the offer, to enter into the contract of insurance, has been transmitted.

§ 20.

Concealment and misrepresentation.

(1) Where there has been, within the meaning of § 19, a concealment or a misrepresentation of a material circumstance, the insurer is, unless otherwise provided, discharged from liability. This rule applies also to the case where a material circumstance has not been disclosed owing to the fact of the person effecting the insurance having had no knowledge thereof through gross negligence.

(2) Where the insurer was aware of the fact which has been concealed or, where he knew a misrepresentation to be wrong, he cannot avoid the contract. The same is true, where the person effecting the insurance has omitted to make a disclosure without any fault of his own.
(3) Where the person effecting the insurance, without any fault of his own, has omitted to make a disclosure, the insurer is entitled to raise the amount of premium due in proportion to the increase of risk (additional premium).

§ 21.

Material circumstance.

Circumstances are deemed to be material—

which the person effecting the insurance has represented as being true, although, in fact, they prove to be wrong—

which the person effecting the insurance has intentionally concealed or misrepresented—

lastly also, as a rule, all such circumstances the insurer has specially enquired into.

§ 22.

Duty of agent to disclose.

The insurer is discharged from liability, where the contract has been concluded through an agent and the agent to insure fails to disclose or misrepresents any material circumstances, which, although unknown to himself, have been known or ought to have been known by his principal.

§ 23.

Alteration of risk.

(1) After insurance is completed, the person effecting the insurance is not allowed, without consent of the insurer, to alter, especially to increase the risk or cause the risk to be altered.
(2) There is an alteration of risk, for instance,—

1. where there has been a considerable delay in commencing or in performing the voyage,
2. where the designated or usual course of the voyage has been departed from, unless the deviation was only a slight one,
3. where, unless it were for the usual call for orders, the ship has called at a port not named in the contract or, where different ports have been designated, the ship has not called in the designated or customary order.

(3) There is an alteration of risk also, where the destination of the voyage has been changed from the destination contemplated by the parties.

§ 24.

Effect of alteration of risk.

(1) Where a loss or damage occurs subsequent to an alteration of risk within the meaning of § 23, the insurer is discharged from liability. Subject to the provisions relating to nondisclosure or misrepresentation of material facts, this rule applies also to the case where the parties have agreed that the risk shall attach at a date prior to the agreement, and the person effecting the insurance has altered the risk or has caused the risk to be altered prior to the agreement but subsequent to the date at which the risk has attached.

(2) An alteration of risk is of no effect,—
when made in furtherance of the interests of the insurer or,
when caused by a peril insured against or,
where the alteration of the risk could in no way have had any influence upon the occurrence of the loss or damage or upon the extent thereof.

§ 25.

Additional premium in case of alteration of risk.

Where the insurer is not discharged from liability on account of an alteration of risk, he may charge an additional premium, if the alteration has been caused by a peril not insured against.

§ 26.

Disclosure of facts increasing the risk.

Unless the insurer is discharged from liability on account of an increase of risk, the person effecting the insurance is bound to disclose to the insurer forthwith any information he receives in connection therewith.

§ 27.

Non disclosure with regard to or alteration of risk of part of the subject-matter insured.

Where there has been with regard to part only of the subject-matter insured a concealment or a misrepresentation of material facts or an alteration of risk discharging the insurer from liability, the insurer is not entitled to claim a discharge from his liability also with regard to the remaining part of the subject-matter insured, unless it can be inferred that he would not have taken the remaining risk for the same consideration. If the insurer thus be partly discharged from his liability, the person effecting the insurance may claim a proportionate return of premium; the insurer, however, is entitled to a consideration in respect thereof within the limits defined in § 18.

SECTION 6.

Perils insured against. Measure of Indemnity.

§ 28.

The risk.

Unless otherwise provided, the insurance is against all perils ship or goods are exposed to during the continuance of the risk. The insurer is specially liable for loss or damage caused by incursion of saltwater, collision of ships, stranding, shipwreck, fire, explosion, lightning, earthquake, ice or by theft, piracy, robbery or other violent acts. The insurer, however, is liable only to the extent defined within these rules; he is not answerable, for instance, for loss or damage caused by a delay of the voyage or for a maritime lien having impaired the value of the thing insured.

§ 29.

General average.

(1) Provided a general average act was performed in order to save ship and cargo from a peril insured against, the insurer is liable for the contributions to be paid by the person effecting the insurance or for the loss caused by the sacrifice of the thing insured. If called upon, the insurer must guarantee the contributions which are to his charge.

(2) In respect of general average sacrifices the same rule applies; where ship and the whole cargo belong to the same owner. The liability of the insurer, in such a case, is to be determined as if ship and cargo were owned by different persons.

§ 30.

General average contributions.

(1) The insurer is liable for the general average contributions as charged in the statement. In the case of § 29, subs. 2, however, the liability of the insurer is determined according to York-Antwerp rules (rule 18 excepted).

(2) The statement, in order to be binding, must be drawn up by persons entrusted therewith by law or local custom.

(3) In default of stipulations to the contrary made beforehand by

the parties concerned, the statement must be drawn up at the place of destination or, in case the voyage has been broken up, at the place where the voyage ends. If there be at that place no average adjuster within the meaning of subs. 2, the average adjuster at the nearest place where such a person is residing must be applied to.

(4) The statement must be drawn up according to the rules of the place where the statement is made, unless the parties have stipulated beforehand that general average shall be subject to York-Antwerp rules or to the rules of the port where the ship is registered. Even if not so provided for by the rules, according to which the statement is made, the insurer is liable for expenses incurred in insuring goods, which have been discharged at a port of refuge, against fire for account of all who are concerned in the general average statement.

(5) The person effecting the insurance must exert himself in safeguarding the interests of the insurer in the drawing up of the statement; the insurer is not liable for any expenses not necessarily incurred in the collection and distribution of the general average.

(6) In spite of a misapplication of the rules, according to which a statement has been drawn up, or a misrepresentation of facts it may contain, the statement is binding upon the insurer, unless the person effecting the insurance can be called to account in respect thereof.

(7) In so far as the person effecting the insurance has been indemnified, the insurer is subrogated to his rights and remedies accruing from the fact that the statement is liable to be disputed. The rules of §§ 45 and 46 apply.

(8) Where less than the full contributing value has been insured, the insurer is liable for such proportion only of the contribution as the sum insured bears to the contributing value.

§ 31.

General average sacrifices.

(1) The liability of the insurer in respect of a general average sacrifice is regulated according to the rules relating to particular average.

(2) The insurer is subrogated to the right of the person effecting the insurance to claim contribution on account of a general average sacrifice immediately the claim has arisen. The insurer, however, who on this account has collected a sum which, after deducting

the expenses incurred in collecting the same, is in excess of the sum he is liable for to pay under the contract of insurance, is bound to remit the balance to the person effecting the insurance. In all other respects the rule of § 45 applies.

§ 32.

Particular charges. Extra charges.

(1) The insurer is liable for and has to make good—

1. all measures reasonably taken and expenses reasonably incurred by the person effecting the insurance in order to avert or minimize an impending loss covered by insurance;
2. all measures taken or expenses incurred at the request of the insurer by the person effecting the insurance at the time the loss or damage occurred;
3. expenses reasonably incurred in ascertaining the loss or damage at the charge of the insurer. Expert's fees and other expenses connected with the help rendered by an expert, an assistant or any other person charged with ascertaining the loss are not allowed, unless such assistance has been rendered upon special request of the insurer or had been stipulated for in the contract of insurance.

(2) The insurer, if asked to do so, has to advance sufficient funds in order to meet the particular charges mentioned in subs. 1, nr. 1 and 2; they remain at his charge, even if they prove to have been inefficient. Where part only of the insurable value has been insured and the parties do not agree as to whether the instructions given by the insurer tend to prevent or minimize the loss, the insurer must advance all sums necessary to meet the expenses to be incurred pursuant to his instructions, although they are to be spent also for the benefit of the part uninsured, and they remain at his charge, if the insurer could not well have supposed his instructions to have been reasonably given and the money has been spent without avail.

§ 33.

Faults committed by the person effecting the insurance.

(1) The insurer is discharged from liability, if the person effecting the insurance wilfully or negligently causes the loss or damage to arise; he is, however, liable for any loss or damage caused by improper navigation (nautical error), provided the person effecting the insurance was not, either intentionally or through gross negligence, privy thereto. Faults or errors committed in shipping, stowing, in the maintenance or in the discharge of the goods do

not come within the meaning of the term "improper navigation".
(2) In an insurance on goods, the insurer is discharged from liability for all loss or damage wilfully or negligently caused by the shipper or consignee of the goods in the discharge of their duties.
(3) The person effecting the insurance is not answerable for acts committed by the crew in the discharge of their duties.

§ 34.

Free of average under 3 per cent.

(1) The insurer is not liable for any loss or damage not amounting to 3 per cent. of the insurable value.
(2) The rule of subs. 1 does not apply—
- to general average contributions or sacrifices,
- to the loss the person effecting the insurance sustains through incurring liabilities to third persons in a case of collision between ships,
- to the expenses referred to in § 32, subs. 1, nr. 1 and 2, and § 95, subs. 3.

Neither these losses or expenses nor the expenses referred to in § 32, subs. 1, nr. 3, are admissible to form part of the amount requisite to constitute a claim within the meaning of subs. 1.
(3) In an insurance on ship, the loss arising upon each voyage is to be treated as a separate average and disallowed if under the required percentage. A voyage within the meaning of this subsection shall commence every time the ship has undergone a new outfitting or when she has been freighted anew or after complete discharge of the cargo or when she sails in ballast to a loading port, the voyage to the loading port being computed as a separate voyage; a voyage shall be deemed to continue until the new voyage begins.

§ 35.

War risks.

(1) The insurer is not liable for loss or damage arising out of war. Thus, he is not answerable for measures taken on account of the war by belligerent Powers, whether these latter have been recognized as such or not; no loss or damage, in particular, is recoverable, which has been caused through the thing insured being stopped, captured, taken, detained, requisitionned, put under restraint or damaged or lost through mines laid on account of the war or through other measures. A measure is considered to have been taken by a belligerent Power, if this Power joins in the war,

raged by another Power, within six months after the measure has been taken.

(2) In an insurance on ship, the insurer is discharged from liability if, on account of the war, the ship calls at a port or does not begin or discontinues the voyage, unless, immediately after having become aware thereof, the person effecting the insurance signifies to the insurer his intention to continue the insurance.

(3) In an insurance on goods, the insurer is discharged from liability, if the goods are discharged on account of the war, unless, immediately after having received information as to the impending discharge or as to the goods having already been discharged, the person effecting the insurance signifies to the insurer his intention to continue the insurance. If the goods remain in port more than two months, the insurer, notwithstanding his liability for other occurrences, is not liable for a subsequent damage or leakage, unless the ship be stranded; the rules of § 114, subs. 1 and 3, apply.

(4) Where there has been a change of the risk insured on account of the war, especially where the ship has called at a port or does not begin or discontinues the voyage, the insurer, unless he is discharged from liability, is entitled to claim an additional premium.

§ 36.

Losses caused by judicial procedure.

Where an order, decree or judgment of a Court has been given or carried out in respect of the subject-matter insured, the insurer is not liable for any loss or damage ensueing therefrom, unless he was liable to make good the claim, on account of which the order, decree or judgment was given.

§ 37.

Measure of indemnity.

(1) The insurer is liable for all loss or damage, arising in the course of the voyage insured, for an amount not exceeding the sum insured. In an insurance on ship, the rules contained in § 34, subs. 3, apply.

(2) Expenses incurred pursuant to the rules contained in § 32 are to be born by the insurer, whether the total amount he thus becomes liable for exceeds the sum insured or not.

(3) Where expenses have already been incurred in order to avert or minimize a loss or for the purpose of ascertaining the loss or damage at the charge of the insurer or for the refitting or the

repairs of the thing insured or where general average contributions have already been paid or the person effecting the insurance has become personally liable on account of them, the insurer is, nevertheless, also answerable for a subsequent loss or damage, even though the total amount of such losses may exceed the sum insured.

(4) The rules contained in subs. 2 and 3, however, do not apply, where and in so far as the total amount to be paid by the insurer exceeds the sum insured by reason of liabilities to third persons incurred in a case of collision between ships for which the insurer has become liable.

§ 38.

Settlement for total loss.

(1) When the insurer settles for a total loss, he is discharged from all further liability.

(2) The insurer, nevertheless, has, in addition to the total loss, to pay all expenses incurred in order to avert or minimize the loss or for the purpose of refitting or repair and all expenses the person effecting the insurance has become personally liable for before the insurer has signified his intention to settle for a total loss. The rule of § 37, subs. 4, applies.

(3) The insurer is debarred from settling for a total loss, if he does not signify his intention to do so within 5 week days after having received information concerning the loss or damage and the immediate consequences thereof.

(4) The insurer, who settles for a total loss, is not vested thereby with any rights in the subject-matter insured.

§ 39.

Insurance on time.

Where the insurance is expressed to be for a certain period of days, weeks or months, the risk commences from noon of the day, at which the contract was concluded, and terminates at noon of the last day of the period fixed in the contract. The time of the place, where the ship happens to be, determines the commencement and the termination of the risk.

SECTION 7.

Duty to inform the Insurer. Duty to avert or minimize a Loss.

§ 40.

The insurer to be kept informed.

The person effecting the insurance is bound forthwith to inform the insurer of all events giving rise to a claim under the policy; he must communicate to him also all accidents of ship or cargo, whether covered or not, materially affecting the risk insured.

§ 41.

Duty to avert or minimize the loss.

(1) In case of accident covered by the insurance, the person effecting the insurance is bound to do his utmost in order to avert or minimize the loss. If he can reasonably do so, he must ask for instructions and, if instructions are given by the insurer, he must carry them out. Where by several insurers instructions have been given inconsistent with each other, the person effecting the insurance must act according to his own best belief and knowledge.

(2) The person effecting the insurance is bound to carry out the instructions, even if part only of the insurable value has been insured. This rule is not applicable to the case, where more than half of the insurable value has been left uninsured; in this case also the rule of the last sentence of § 32, subs. 2, as to the liability for and advances to be made on account of intended expenditures does not apply.

(3) Unless there has been no fault or negligence on the part of the person effecting the insurance in not or not properly complying with his duty to avert or minimize the loss, the insurer is not liable for any loss or damage arising thereout.

SECTION 8.

Loss must be notified. Evidence to be produced. Claim when due.

§ 42.

Loss must be notified.

(1) The person effecting the insurance must notify in writing a loss, in respect of which he desires to make a claim, to the insurer within 15 months to be counted from the termination of the risk or, in case of a missing ship, from the date at which, according to the rule contained in § 72, subs. 1, the ship is presumed to be lost. In order to comply with this rule, it is sufficient that the notice be sent off, though it be not communicated within the time prescribed.

(2) The person effecting the insurance is debarred from making a claim, if the loss be not notified within the time prescribed in subs. 1.

(3) These rules do not apply to a claim for general average

contributions for which the person effecting the insurance has become liable.

§ 43.

Evidence of loss.

After an accident has happened giving rise to a claim under the contract of insurance, the insurer has a right to be informed by the person effecting the insurance of everything required for ascertaining the loss or the amount of the liability of the insurer. In so far as the person effecting the insurance can reasonably be expected to be able to procure documents, documentary evidence must be produced at the request of the insurer; also a sea protest can be called for, if the insurer shows sufficient reasons for this request.

§ 44.

Presentation of account. Claim when due.

(1) The person effecting the insurance, after having presented his account and having produced all documents required by the insurer, must wait a month before he can ask for payment. If he has not presented his account or produced the evidence required, without fault or negligence on his part, within a month from the day at which he has notified the loss to the insurer, he is entitled to a payment on account of the sum total of his claim of three-fourths of the amount which, judging from the present state of affairs, the insurer is at least liable for.

(2) The account must show correctly arranged the different amounts the insurer shall pay for each single loss or damage and expenditure. A particular average statement must be drawn up, at the request of the insurer, by persons entrusted therewith by law or local custom at the place where the account shall be settled.

SECTION 9.

Subrogation.

§ 45.

Subrogation.

(1) The right of the person effecting the insurance to claim damages from a third person passes to the insurer, to the extent to which these damages have been paid by the insurer. The person effecting the insurance is bound to give all information necessary for the prosecution of the claim and to deliver to the insurer all documents in his possession by which his claim can be established. If asked to do so, he must furnish the insurer, at the latter's expense, with an authenticated title as to the cession of his claim.

(2) To the extent to which the insurer would have been in a position to indemnify himself out of the rights ceded to him in accordance with the rule of subs. 1, he is discharged from liability towards the person effecting the insurance, if the latter waives his right to recover damages from a third person or a right or title securing this claim.

§ 46.

Duty to minimize the loss notwithstanding subrogation.

The subrogation of the insurer to the rights and remedies of the person effecting the insurance does not relieve the latter from his duty to make exertions in order to minimize the loss, for instance by staying the payment of freight. He must inform the insurer immediately upon receiving any news of importance for the prosecution of the claim and must render, if asked to do so, every assistance necessary for that purpose, especially by sueing in his own name. The costs of such a proceeding are to be born by the insurer; he must, if requested to do so. advance the sums necessary to meet the expenses.

SECTION 10.

Insolvency of Insurer.

§ 47.

When the insurer turns insolvent and does not tender sufficient securities to meet the event, the person effecting the insurance is entitled to cancel the contract and to insure anew at the expense of the insurer.

SECTION 11.

Statutary Limitations.

§ 48.

All claims arising out of a contract of insurance become prescribed within 5 years. The period of limitation does not begin before the year has elapsed in the course of which the risk has terminated or, in the case of a missing ship, the ship is presumed to be lost according to the rule contained in § 72, subs. 1.

SECTION 12.

Transfer of Thing insured.

§ 49.

General principle.

(1) Where the property in the thing insured has been transferred, all rights and liabilities arising out of the contract of insurance

subsequent to the transfer pass from the person effecting the insurance and are vested in and incumbent on the assignee; the assignor, however, and the assignee are both conjointly liable for the payment of the premium.

In an insurance on goods, provided a policy has been issued, the assignee is not liable for the payment of the premium including additional charges; nor can the insurer, in such a case, claim to be discharged from his liability towards the assignee in virtue of the rule contained in § 17, unless the assignee knew or ought to have known that the premium has not been paid or security been tendered as provided for in § 16, subs. 3.

(2) Until the insurer has been informed of the transfer, the transfer does not produce any effect upon the relations existing between the insurer and the person effecting the insurance. As to the relations between the insurer and the assignee, the rules of §§ 406 to 408 German Civil Code apply.*)

In an insurance on goods, provided a policy has been issued, these rules do not apply, unless the assignee knew or ought to have known that the premium has not been paid or security been tendered as provided for in § 16, subs. 3.

---

*) § 406. The debtor (insurer) is not debarred from making a set off against the assignee with a claim he has against the assignor, unless he was aware of the assignment when the claim accrued or, although he was unaware of the assignment when the claim accrued, he was informed thereof and the debt assigned had fallen due prior to the date at which the claim matured.

§ 407. The debtor (insurer) is entitled to set up against the assignee any payment or transaction with regard to the debt assigned, which, though subsequent to the assignment, has been operated or been agreed upon between the assignor and himself before he was informed of the assignment.

If, after the assignment, an action arose out of and final judgment was delivered with respect to the original contract between the debtor and the assignor, the debtor is entitled to set up the judgment against the assignee, unless he was informed of the assignment when the action arose.

§ 408. In case of successive assignments the rules of §§ 406 and 407 apply to the objections the debtor (insurer) is entitled to raise against the claim brought against him by one of the assignees and which he derives from payment or any other transaction or judicial proceedings, which have passed between him and any other assignee in respect of the original contract, or which consist in a set off the debtor intends to make with a claim he has against the other assignee.

This rule applies also to the case, where, after the debt has been assigned by agreement, a third person has acquired the same by operation of law and has received from the assignor the acknowledgment of this fact, or where the debt has been transferred to a third person by judicial decree.

(3) The insurer is not liable for any loss or damage which would not have occurred but for the transfer of the thing insured. This rule, however, does not apply to an insurance on goods, unless the property in the goods has been transferred to a subject of a belligerent Power in the course of a war.
(4) The assignee is entitled to cancel the contract of insurance without warning within a month after the assignment or, having had no knowledge of an insurance having been effected, within a month after he has been informed thereof. If the assignee cancels the contract, he is not liable for the premium.
(5) The subsections 1—4 apply likewise to the case of a forced sale of the thing insured.

§ 50.

Transfer of ships and ship's shares.

(1) The rules of § 49 apply in the case of a transfer of ship's shares.
(2) When the property in a ship has been transferred, the contract of insurance is dissolved; the premium, however, is apportioned over the time before and subsequent to the transfer; the insurer is entitled to receive the proportionate part of the premium due for the former and one fifth of the proportionate part due for the latter period.

Where the property in a ship has been transferred in the course of a voyage, the insurance terminates at the next port of destination subject to the rules laid down in §§ 66 to 68.

§ 51.

Mortgage of claim against insurer.

Where the claim against the insurer has been mortgaged, the insurer cannot claim to be discharged from his liability towards the mortgagee in accordance with the rule contained in § 17, unless the mortgagee knew or ought to have known that the premium has not been paid or security been tendered as provided for in § 16, subs. 3.

SECTION 13.

Insurance for account of another.

§ 52.

Insurance for one's own account, for account of another, for account of whom it may concern.

(1) Where it does not appear that the person effecting the insurance intends taking the insurance, though in his own name, but on

behalf of another (insurance for account of another), the insurance is, nevertheless, held to have been taken for account of the person effecting the insurance (insurance for one's own account).

(2) Where an insurance has been taken on behalf of another, the person effecting the insurance, even where the name of the other person has been disclosed, is presumed to have taken the insurance in his own name, though for account of the other person.

(3) Where an insurance has been taken for account of whom it may concern or, where it appears from the contract that it shall be left undetermined whether the insurance has been taken in the own interest of the person effecting the insurance or in the interest of another, the rules relating to insurances for account of another apply when it appears that the interest of another person has been insured.

§ 53.

Position of assured.

(1) The rights arising out of the contract of insurance are vested in the assured. The policy, however, where it has been agreed that a policy shall be issued, must be handed to the person effecting the insurance.

(2) Without the assent of the person effecting the insurance, the assured can not use or dispose of his rights or begin legal proceedings, unless he is in possession of the policy.

§ 54.

Position of person effecting the insurance.

(1) The person effecting the insurance can dispose in his own name of the rights arising out of the contract of insurance and which are vested in the assured.

(2) Where a policy has been issued, the person effecting the insurance is not entitled, without the assent of the assured, to accept payment or to assign the rights of the assured, unless he is in possession of the policy.

(3) The person effecting the insurance is not entitled to ask the insurer to make payments to him, unless he proves that the assured has assented to the insurance being taken.

§ 55.

Assured and person effecting the insurance.

The person effecting the insurance is not bound to deliver the policy to the assured or, in case he became bankrupt, to the trustee, until he has been satisfied for all claims he has against the assured in respect of the subject-matter insured. In respect of these claims he is entitled to satisfy himself out of the claim against the insurer or out of the sum the insurer has paid, in priority over the assured and the other creditors of the assured.

§ 56.

Set off by insurer.

The insurer is entitled to set off against what he owes on account of the contract of insurance any claim he has against the person effecting the insurance in so far as the claim arises out of the insurance which has been taken on behalf of the assured.

§ 57.

Knowledge of person effecting the insurance and of assured as to facts to be communicated to insurer.

(1) With regard to the nondisclosure or misrepresentation of material facts, the person effecting the insurance is deemed to know or ought to know all facts the assured knows or ought to have known. The same rule applies to the case where the person effecting the insurance is discharged from his liability to pay the premium on account of the interest not having attached (§ 4).

Where the question is raised of avoiding the contract on account of the nondisclosure of a material fact, it is not sufficient. in order to set up a valid defence, to show that the omission of disclosing the fact was due to no fault of the person effecting the insurance (§ 20, subs. 2), but it must also be shown that the omission was due to no fault of the assured.

(2) Where the risk shall attach at a date prior to the conclusion of the contract of insurance (§ 5), the insurer is discharged from liability not only, if the person effecting the insurance, but also if the assured himself knew or ought to have known, at the time the agreement was made, that the event insured against had already occurred.

(3) The fact that the assured had known or ought to have known a circumstance which should have been communicated to the insurer, is of no consequence in case the contract was entered into without his knowledge or where he could not have informed the person effecting the insurance in due course. The communication is not deemed to have been made in due course, if it has not been transmitted as speedily as is consistent with the ordinary course of business or, at least, in the same or a similar mode to

which the order to insure has been sent to the person effecting the insurance.
(4) The person effecting the insurance can not set up the plea that the insurance was taken without the knowledge of the assured, if he did not inform the insurer, at the time he made the agreement, that he did so without the authority of the assured.

SECOND PART.
Special Rules on special Subjects.

SECTION 1.
Insurance on Ship.

§ 58.
Seaworthiness.

(1) The insurer is not liable for loss or damage arising out of the unseaworthiness of the ship, especially not for any loss or damage caused through the ship having put to sea improperly equipped, manned or documented or being unseaworthy on account of the nature or the stowage of the cargo. The insurer is not liable for any loss or damage arising before the ship has put to sea and caused through the ship not having been in a state fit to encounter the dangers of the port.
(2) Where loss or damage have occurred not attributable to any external event, the loss is presumed to have arisen out of the unseaworthy condition of the ship as explained in subs. 1.

§ 59.
Wear and tear.

The insurer is not liable for any loss or damage the ship sustains owing solely to wear and tear, to decay on account of age, rottenness, rust or the action of worms. Thus, where sails are carried off or split or are otherwise damaged, even if it be through carrying press of sail, or where anchors, ropes, chains or the rigging has been damaged or, on account of such a damage, ropes or sails have been cut away, or the anchors, anchor chains or -cables had to be cut away or slipped, it is accounted to be a damage owing to wear and tear. But, where sails have been injured by blows of the sea or by the breaking of spars or where damage has been done to sails while set, this is not a loss by wear and tear.

§ 60.

Dangerous goods.

(1) Where explosives or goods, which are liable to spontaneous combustion, have been shipped in contravention of the rules issued by the German States bordering on the sea, and which relate to the shipment of dangerous goods in merchant vessels, or where the conditions under which, according to these rules, the shipment of certain dangerous goods is allowed, have not been complied with, the insurer is not liable for any loss or damage ensueing therefrom. The same rule applies, where goods have been shipped at a foreign port and the regulations there in force have not been complied with. The insurer, however, is liable, where the person effecting the insurance neither knew nor ought to have known that such a shipment had taken place or that, in shipping the goods, the regulations have not been complied with; the insurer, in such a case, is entitled to demand an additional premium.
(2) Where more than a third part of the carrying capacity of a ship has been loaded with bones, bone-ash, cement, cereals or other produce shipped in bulk, charcoal, chaulk, clay, residue of clay, coal, earth, hay, iron (especially rails or iron girders), lime, manure, mineral (especially marble, slate, stones or tiles) or salt, the insurer is not liable for any loss or damage arising thereout, unless the person effecting the insurance did not know nor ought to have known that such cargo or such a quantity of the said goods had been shipped. If the insurer is not discharged from his liability, he is entitled to demand an additional premium.

§ 61.

Breaking through ice.

The insurer is not liable for any loss or damage arising out of the ship breaking through blocked ice, unless it was necessary in order to avert or minimize a loss covered by the insurance.

§ 62.

Sacrifice of deck cargo.

The insurer is not liable for general average contributions on account of jettison of deck cargo.

§ 63.

Sacrifice where ship sails in ballast.

Where a ship sails in ballast and sacrifices have been made in the course of the voyage, which would have been allowed in general average if cargo had been on board, the insurer is not liable, unless he was informed, at the time the contract was made, that the ship intended to sail or already sailed in ballast.

§ 64.
Reward for services rendered to be taken in account.

Damage done to a ship whilst employed in lightering or towing another vessel or in the act of salvage, is not recoverable in so far as the reward stipulated or due for the assistance rendered serves to indemnify the shipowner for any loss or damage sustained.

§ 65.
Damage done to cabin or machinery.

No damage is recoverable done to stock or fittings of the cabin or to the machinery, unless the ship be stranded. The term "machinery" includes in particular, the principal engine (including donkey engines or other contrivances employed in working the principal engine, the boiler and funnel, the line of shafting, propeller and wheel), all engines on deck, steering engines and the part of the steering chain connecting the engine with the steering quadrant, pumps for ship's use, appliances for electric light and wireless telegraph, refrigerating machines. As to the meaning of the clause "unless the ship be stranded" vide infra § 114, subs. 1 and 2.

§ 66.
Duration of risk.

(1) Where a ship has been insured for a specified voyage, the risk commences immediately the ship begins with taking in the cargo or ballast or, where no goods or ballast are to be shipped, upon the sailing of the vessel. The risk terminates, when the discharge of the cargo or ballast at the port of destination has been completed or, where no cargo or ballast is to be discharged, immediately the ship has moored at anchor at her proper place. If the discharge is unduly delayed by the person effecting the insurance, the risk terminates when the discharge would have been completed without the delay.

(2) Where, before the discharge is completed, cargo or ballast has been taken in for a new voyage, the risk terminates when the new loading begins.

(3) Where the voyage has been abandoned, after the risk has attached, the insurance terminates at the place where the voyage

ends, in accordance with the rules relating to the termination of the risk at the port of destination.

§ 67.
Continuation of risk.

(1) If the ship arrives at her port of destination in a damaged condition and the insurer is liable for the damage sustained, the risk continues until the ship has been repaired. If the repairs are unduly delayed by the person effecting the insurance, the risk terminates when the repairs would have been completed if no delay had taken place. If, immediately upon being informed of the damage, the person effecting the insurance gives notice to the insurer that the risk shall not continue to run, the risk terminates as provided for in § 66. In case the risk is continued, the insurer is entitled to demand an additional premium in proportion to the time the risk continues to run.
(2) The rules of subs. 1 do not apply where repairs have to be effected at a place other than the port of destination and the question, whether the ship is able to reach the port of repairs, has not been ascertained in the manner prescribed in § 74.

§ 68.
Continuation of risk in insurances on time.

Where a ship has been insured for a definite period of time and is still at sea when the period expires, the risk continues to run until the ship has reached the next port of destination, and terminates there in the manner prescribed in § 66. The risk does not continue to run, if, before the ship has started on her last voyage, the person effecting the insurance has given notice to the insurer that the risk shall terminate at the time provided for in the contract. In case the risk is continued, the insurer is entitled to demand an additional premium in proportion to the time the risk continues to run; in case the ship is missing, the amount of premium to be paid must be calculated up to the time at which the ship is presumed to be lost (§ 72. subs. 1).

§ 69.
Successive voyages insured in one contract.

Where a ship has been insured for several successive voyages, the whole adventure is to be considered as one voyage and the insurer is responsible for any loss or damage that may happen in the whole course of its duration.

§ 70.

Insurable value.

(1) In insurances on ship, the insurable value is the value of the ship at the commencement of the risk, not including her outfit, seamen's wages or the charges of insurance.
(2) The amount to be recovered in case of loss or damage is to be fixed upon the value as ascertained according to the rule of subs. 1.

§ 71.

Total loss.

(1) In case of total loss, the measure of indemnity is the sum insured, deduction being made of the value of what has been saved of the ship, before the insurer has settled for a total loss, and of all sums the person effecting the insurance has received on account of the loss. Upon the request of the insurer, the value of what has been saved must be ascertained by public auction.
(2) There is a case of total loss also where the person effecting the insurance has been deprived of the ship without hope of recovery, especially where the ship has foundered without the possibility of being salved, or where the ship has been reduced to such a state as to be no longer capable of use under the original denomination.
(3) Where the insurer has paid for a total loss, he is subrogated to all rights and remedies of the person effecting the insurance in and in respect of the ship. The person effecting the insurance is bound to give all information necessary for the prosecution of his rights and to hand over to the insurer all documents in his possession by which his rights can be established. If asked to do so, he must furnish the insurer, at the latter's expense, with an authenticated title as to the cession of his rights.
(4) The subrogation of the insurer to the rights and remedies of the person effecting the insurance does not relieve the latter from his duty to make exertions in order to minimize the loss, in so far as the insurer is not capable of doing so himself. He must inform the insurer immediately upon receiving any news of importance for the prosecution of his rights and must render, if asked to do so, every assistance necessary for the recovery of the ship and for the purpose of turning into account whatever remains of her. The costs are to be born by the insurer; he must, if requested to do so, advance the sums necessary in order to meet the expense.

§ 72.

Missing ships.

(1) The person effecting the insurance is entitled to ask for a settlement of a total loss where a ship is missing, that is to say: where the ship has not arrived at her next port of destination nor any news have been received from her within or for at least three times the time she would have employed, under ordinary circumstances, to sail from the place, from which the last news was dated, to her next port of destination; steamers, however, must have been missing for at least 2 months and sailing vessels for at least 3 months before they are presumed to be lost. Where the forwarding of news might possibly have been delayed on account of war, instead of 2 or 3 months, ships must have been missing 6 months before they are presumed to be lost.
(2) The person effecting the insurance cannot make his demand for a settlement of a total loss in case of a missing ship depend upon conditions or upon the expiration of a certain term.
(3) The insurer, upon receiving the notice aforesaid from the person effecting the insurance, is subrogated to all his rights and remedies in and in respect of the ship (abandonment). The rules of § 71, subs. 3 and 4, apply.
(4) The person effecting the insurance is debarred from asking to be paid as for a total loss if, before doing so, news concerning the ship have been received.

§ 73.

Seizure and detention. Pirates.

The rules contained in § 72 apply to the case where a ship has been seized or detained by authority or has been taken by pirates. Two months, however, must in any case elapse, before a settlement as for a total loss can be asked for.

§ 74.

Partial loss.

(1) All partial losses must be ascertained by experts.
(2) Two experts must be immediately appointed, one by the insurer and the other by the person effecting the insurance or, on his behalf, by the master.
(3) If the two experts disagree, they call an umpire; the umpire may also be appointed beforehand. If they disagree as to who shall be appointed as umpire, each of them designates another

expert and lots are drawn as to which of these two shall be umpire. If the ship lies in a foreign port and the two experts cannot arrive at an agreement, the person effecting the insurance or, on his behalf, the master shall, for the purpose of nominating an umpire, apply to the consul of the country, in which the ship is registered or, where there is no such consul or where the consul is not willing to appoint an umpire, to the consul of another country or, where there is no such consul either or where the consul of the other country is also not willing to comply with the request, to an official of the country, in which the ship is, authorised to appoint experts.

(4) If the insurer, although requested to do so, does not appoint an expert, the person effecting the insurance or, on his behalf, the master may apply to the chamber of commerce, in whose jurisdiction the ship is lying, for the appointment of an expert. If the ship lies in a foreign port, the expert must be appointed according to the rules of subs. 3 third sentence.

(5) The experts take a survey, examine the damage and make a report. As far as possible, all persons concerned must take part in the survey. The report must contain:—

1. Name of experts and persons present at the survey.
2. Name of persons, by whom experts have been appointed.
3. Place and time at which the survey has been taken and the damage has been ascertained.
4. Each single item of damage and, as far as possible, the cause, especially whether a loss or damage has been caused through perils of the sea on the ship's last voyage or whether other reasons must be accounted for.
5. Estimate of costs of repairing each single loss or damage through perils of the sea during the last voyage of the ship.

(6) The decisions must be taken by majority of votes. If for certain sums no majority can be obtained, a majority will be arrived at by adding the vote for the biggest sum to the vote for the next biggest amount.

(7) The report must be signed by the experts and, if present, by the agent of the insurer, in order to show that he has taken part in the survey.

(8) The decision of the experts is binding, unless it be openly inconsistent with the actual state of affairs. In this last case, the damage must be ascertained by proceedings in Court. If the experts are not willing or able to ascertain the damage or if they

are guilty of undue delay, other experts must be appointed in accordance with the rules of subss. 2 to 4.

(9) The insurer may refuse payment, until the loss or damage has been ascertained in accordance with the foregoing rules. If, without fault or negligence of the person effecting the insurance, the damage has not been ascertained in the manner aforesaid, the insurer may refuse payment, until the loss or damage has been ascertained in some other suitable way.

(10) The claim's agent of the insurer, appointed for the district in question, is authorised to accept all statements made to him by the person effecting the insurance with regard to the adjustment of the partial loss, and to deal with him and enter into transactions on this account in the name of his principal.

§ 75.

Repairs.

(1) Immediately the partial loss has been ascertained, the ship is to be repaired. As regards the repairs, the person effecting the insurance has also to look after the interests of the insurer. Before the contract for repairing the ship is concluded, the insurer, if possible, must be consulted and a draft of the agreement must be shown to him.

(2) The insurer is entitled to superintend the repairs. If asked to do so, the person effecting the insurance must keep the insurer informed of the progress of the repairs. After the repairs have been completed, he must produce the invoices and show the insurer, in particular, whether any abatements or deductions have been made.

(3) The person effecting the insurance is entitled to the actual cost of repairs, but not exceeding the sum estimated by the experts, deduction being made —

1. first of an amount on account of new for old (§ 76).
2. secondly of the value of all old materials which have been replaced by new ones. Upon request of the insurer, the value of these materials must be ascertained by public auction.

(4) The person effecting the insurance, moreover, is entitled to the expenses necessarily incurred in bringing the ship to her place of repairs and, eventually, back again, and also to the expenses incurred in raising the sums necessary for repairing the ship.

(5) If the person effecting the insurance can show special reasons which he is personally not answerable for, for not repairing the

ship, he must inform the insurer thereof immediately after the damage has been ascertained; the fact of having transferred the property in the ship, before repairs have begun, without assigning the contract of insurance to the transferee, is considered a good reason for not repairing the ship. The person effecting the insurance is, in such a case, entitled to the estimated cost of repairs, less deductions on account of new for old and a reasonable sum on account of new materials, which would have been substituted for old ones if the repairs had been effected.

(6) Any differences as to whether or not the person effecting the insurance has sufficient reasons for not repairing the ship must be settled by arbitration. The person effecting the insurance and the insurer each appoint an arbitrator. If the arbitrators do not agree, they appoint an umpire or, if they do not agree as to the person to be nominated as umpire, the umpire is chosen by the chamber of commerce or any other body representing the interests of the commercial community at the place of business of the insurer or his agent, whereever the contract of insurance has been concluded.

§ 76.

Deductions new for old.

(1) Deductions new for old are made in accordance with the rules laid down in subss. 2 to 7.

(2) Unless otherwise provided, the deduction is of one-third new for old.

(3) No deduction is made on account of ironwork or cementing of iron ships, provided the loss or damage occurred during the first ten years from date of original register. If the loss or damage occurred in the course of the next 5 years, a deduction of one-sixth is made, later on of one-third. Where damaged parts can be lashed or straightened, this must be done, unless renewing the parts is necessary for retaining the ship's class; costs of lashing or straightening are allowed in full. From the costs of painting or scraping the bottom, one-third is deducted; no painting or scraping to be allowed, if the bottom has not been painted or scraped within six months previous to the date of accident. With regard to the wooden parts of iron ships, the rules relating to wooden ships apply.

(4) As regards wooden or composite ships no deduction is made from the costs of caulking or metal sheathing, if the loss or damage occurred within the first year after the ship has been

caulked or careened. If loss or damage occurs in the course of the second year, a deduction of one-third is made, if in the course of the third year, of two-thirds; no caulking or metal sheathing to be allowed later on. Costs of repairs to or renewal of hull or masts are allowed in full, if the loss or damage occurs within the first year from the launching of the ship. Likewise repairs to or renewal of detached parts of the hull or the masts or of one or the other appurtenance are allowed in full, if the object to be repaired or to be renewed was new when the ship started on the voyage, in the course of which the loss or damage occurred.

(5) Costs of repairs to or renewal of machinery are allowed in full, if the loss or damage occurs within the first three years from the date at which the particular part was put in use. If the loss or damage occurs within the next three years, a deduction of one-sixth, afterwards a deduction of one-third is made.

(6) With regard to boilers, no deduction is made, if the loss or damage occurs within the first year after the boiler has been put in use; for every following year in which the boiler has been used, one-tenth is to be deducted. No costs are allowed after the boiler has been used for ten years.

(7) Anchors are allowed in full. Anchor chains are allowed in full, if the loss or damage occurs within the first year after they have been put in use: otherwise they are subject to a deduction of one-sixth.

(8) Unless otherwise provided, no costs are allowed for painting or for repairs to or renewal of glassware.

§ 77.

Constructive total loss.

(1) Where a ship, through perils insured against, has been damaged to such an extent, that she cannot be repaired and this fact has been ascertained in the manner prescribed in § 74, the person effecting the insurance is entitled to sell the ship by public auction and to claim the difference between the sum insured and the proceeds of the sale. A ship is considered as being unfit for repairs, where either the ship is not capable of being repaired or where repairs cannot be effected at the place, where the ship is, and the ship cannot be brought to a place, where she could be repaired. Deduction must be made of all sums, the person effecting the insurance receives from other persons on account of his loss,

and of the amount of all loss or damage arising out of perils not insured against. The risk continues to run, until the ship has been sold; the insurer is entitled to demand an additional premium in proportion to the time the risk continues to run.

(2) The rule of subs. 1 applies to the case of a ship having been damaged, through perils insured against, to such an extent that she is not worth being repaired, provided this fact has been ascertained in the manner prescribed in § 74. A ship is considered as not being worth being repaired, when the costs of repairing her, estimated according to the rules contained in § 74, without considering deductions on account of new for old, do exceed her insurable value.

(3) The rule of subs. 1, the last sentence excepted, applies also in a case, where the fact of the ship not being capable of or not being worth being repaired has become apparent at a later stage, for instance during the progress of the repairs, and the fact has been ascertained forthwith in the manner prescribed in § 74. The insurer has to pay the costs of repairs in so far as the ship, on account of the repairs, has been sold for a higher price.

(4) The person effecting the insurance is debarred from selling the ship by public auction and claiming the difference between the sum insured and the proceeds of the sale, if he does not notify his intention to do so immediately after the facts, which justify his proceeding in the manner aforesaid, have been established and he has been informed thereof.

§ 78.

Liability to third parties in case of collision.

(1) The insurer is liable for all loss or damage the person effecting the insurance sustains through incurring liabilities to third persons in a case of collision between ships.

(2) The insurer is liable for such proportion of the measure of indemnity as the value of the ship bears to the value of the ship plus the amount of freight.

§ 79.

Other insurances with regard to the ship.

Unless otherwise provided, the rules of section 1 apply also to other insurances covering any interest in or with respect to a ship.

SECTION 2.
Insurance on Goods.

§ 80.
Meaning of "goods and merchandises".

In an insurance on any kind of goods and merchandises the following goods are not included:—

1. gold, silver, platina, specie, bank notes, bills of exchange, securities, coin, precious stones, objects made of precious metal, pearls, real lace, works of art.
2. objects liable to explosion or spontaneous combustion (explosives, ammunition, fuse, pyrotechnical goods, compressed and liquid gases developping if brought into contact with water inflammable or other gases promoting combustion), petroleum, naphtha, benzine and quick lime.
3. goods of the kind of those mentioned in § 60, subs. 2, if exceeding a third part of the carrying capacity of the ship, unless the person effecting the insurance has not given his consent to the shipment. The insurer, in this last case, is entitled to claim an additional premium.

§ 81.
Goods disposed of in order to prosecute the voyage.

The insurer is liable if, in order to prosecute the voyage, the goods have been disposed of either in exchange for a respondentia bond or in any other way.

§ 82.
Liability of insurer for damage sustained.

(1) As a general rule, the insurer is not liable for damage done to the goods or for the death of live stock, unless the ship be stranded. As to the meaning of the clause "unless the ship be stranded" vide infra § 114, subss. 1 and 3.

(2) The insurer is liable, whether there is a case of stranding or not,—

1. where the following goods have been insured: arsenic, asbestos-ware, asphalt, borax in casks, braids, buttons, camphor raw, caoutchouc-ware, cardamom in cases, carpets, celluloid-ware, cinnabar, cinnamon, cloths, clothings ready made, coffee, copperwire insulated, cotton, cottons, elastics, elephants tusks, embroidery, estamine, felt-ware, gold-ware, gum copal, half woollen goods, horns, horn-tips, hosiery, India rubber, India rubber goods, genuine indigo, jute-ware [bags and baggings (Hessians) excepted], knitting wool, lac-dye, laces, dressed leather (patent leather excepted), leather goods

manufactured, linen, linens, lingerie or other white linen, linoleum, mace in casks or cases, manufactured goods not specially mentioned, mercury in metal vessels, metal pigs slabs or bars (iron and steel excepted), musk, nickel, nutmegs, opium, pepper, pitch, resin in casks, ribbons, tarred rope, saddle-cloth, shellack, silk, silk goods, silver-ware, genuine silver wire, spermaceti, stearine, sulphur crude, thread, tortoise-shell, trimmings not specially mentioned, umbrellas, vanilla, wax (raw wax and wax combs excepted), wool, woollens, yarn (including Turkish red yarn and spool yarn), zinc plates.

2. where it has been specially agreed that the insurer is not liable for any loss or damage not amounting to or not exceeding a certain percentage of the insurable value, or that he is liable only if the loss or damage amounts to or exceeds a certain percentage of the insurable value.

§ 83.

Goods transshipped and return goods.

(1) The insurer, notwithstanding his liability for other occurrences, is not liable for damage done to the goods, unless the ship be stranded, where the goods have been conveyed by sea or inland waters to the place, where they are going to be or have been transshipped into the vessel, which is going to take them on the voyage covered by the insurance, unless the insurer has been informed of the previous voyage at the time the contract was concluded or the damage sustained could not possibly have occurred but on the voyage covered by the insurance.

(2) The rule of subs. 1 applies, likewise, in a case where the goods, after having reached their place of destination, have been unpacked (wholly or in part), have been repacked, because they have been found to be unsalable or for other reasons, and are going to be or have been sent on or sent back upon the voyage covered by the insurance—especially, where the goods are return goods—, unless the insurer was informed thereof at the time the contract was concluded or the damage sustained could not possibly have occurred but on the voyage covered by the insurance.

§ 84.

Goods already damaged.

Where the goods have been damaged before they start on the voyage insured, the insurance is free of all damage sustained, unless the insurer has been informed thereof at the time the contract was concluded. As to the liability of the insurer, the rule of § 113 applies. Where the goods have been lost, the insurer is not liable beyond the actual insurable value.

§ 85.

Deck cargo.

(1) Where goods have been stowed on deck, the insurer is only liable for general average contributions chargeable to the person effecting the insurance and for a loss of the goods through one of the events mentioned in § 73 or through a total loss of the ship or on account of the ship being considered a total loss in accordance with the rules contained in § 72.
(2) The rule of subs. 1 does not apply, where the goods have been stowed on deck without the consent of the person effecting the insurance. The insurer, in such a case, is entitled to claim an additional premium.

§ 86.

Inherent vice.

(1) The insurer is not liable for loss or damage caused by the nature of the goods, especially by inherent vice, wasting, rust, mould, ordinary leakage and breakage, spontaneous combustion, or by improper packing, sweat, rats or mice. Loss of liquid goods to the extent of 5 per cent., if in metal flasks up to 3 per cent. is considered to be within the limits of ordinary leakage.
(2) The insurer, notwithstanding his liability for other occurrences, is not liable for extraordinary leakage nor for any damage done to liquid goods or their packing or fittings, unless the ship be stranded. As to the meaning of the clause "unless the ship be stranded" vide infra § 114, subss. 1 and 3; there is a case of stranding, however, also where, on account of an accident by perils insured against, the ship has called at a port of refuge and the goods have been discharged.

§ 87.

Clauses exempting carrier from liability.

The insurer is not liable in so far as the person effecting the insurance is debarred from claiming compensation on account of the carrier of the goods having exonerated himself beyond the legal or customary limits.

§ 88.

Duration of risk.

(1) The insurance covers the whole extent of the voyage insured.
(2) The risk commences when the goods have been delivered to the carrier either for immediate shipment or to be taken charge of until they can be shipped. Where the goods have been delivered by the person effecting the insurance to a wharf or warehouse, the wharfinger or warehouseman is considered to have received the goods as agent of the carrier. The carrier has taken charge of the goods within the meaning of this section only in case he has done so for a short time not exceeding the customary limit.
(3) The risk terminates when the goods have been delivered to the consignee at the place of destination or, where they have not been so delivered, when they have been duly deposited or sold, but not later than on the tenth day after complete discharge. Where the discharge has been unduly delayed by the person effecting the insurance, by the shipper or consignee, the risk terminates on the tenth day from the date at which the discharge would have been completed if no delay had taken place.

§ 89.

Risk of craft.

The insurance covers such craft risk as is usual for the purpose of loading or landing the goods at the particular port.

§ 90.

Insurable value.

(1) The insurable value of the goods is the value they have in trade or, if there is no such special value, the common price, current at the place of shipment at the time the risk has commenced according to the rules of §§ 88 and 89, plus the charges of insurance, the expenses incurred up to the time the goods have been delivered to the carrier and the freight paid independently of delivery.
(2) The amount to be recovered in case of loss or damage is to be fixed upon the value as ascertained according to the rule of subs. 1.

§ 91.

Total loss. Abandonment.

(1) In case of total loss the rules of § 71 apply. Where, however, the goods have only been reduced to such a state as to be no

longer capable of use under the original denomination, this fact must be ascertained by experts in accordance with the rules of § 74, subss. 2 to 10, if the person effecting the insurance wants to claim for a total loss.

(2) Where the goods have been seized or detained by authority or have been taken by pirates or, where the ship is considered to be totally lost within the meaning of § 72, the rules of §§ 72 and 73 apply.

§ 92.

Total loss of part.

The rules for adjusting a total loss apply also to the case of a total loss of part.

§ 93.

Damage.

(1) In case of damage done to goods, the current value in trade or, if no such value exists, the common price must be ascertained, for which the goods would have sold at the place of destination, had they arrived there sound (sound value), and also the value or price for which they sell arriving there damaged (damaged value). The measure of indemnity is such proportion of the insurable value as the difference between the sound and damaged value bears to the sound value of the goods.

(2) The damage done to the goods, their sound and their damaged value must be ascertained by experts in accordance with the rules contained in § 74, subss. 2 to 10. The estimate must include all public dues, especially the custom duties. Where goods have arrived some sound and some damaged or, where parts of a parcel have been lost or damaged, the damaged goods must, if practicable, be separated from the sound ones; this rule applies in particular to goods which have been packed together.

(3) Upon the request of the insurer, the damaged value must be ascertained by public auction, unless the amount of public dues to be paid by the damaged goods at the place of destination exceeds their damaged value, as ascertained by the experts, exclusive of public dues; the insurer, however, is debarred from requiring the goods to be sold by public auction, if he does not notify his request within a week after the experts have made their report. If the goods are sold by public auction, the gross proceeds of the sale take the place of the damaged value for the purpose of adjustment. If by the rules and regulations of the auction the seller has to advance money in order to cover the expenses thereof, the insurer,

provided he assented to the rules and regulations aforesaid, is answerable for the purchase-money being paid.

(4) Until the damage has been ascertained, the person effecting the insurance is not allowed, without the consent of the insurer, to make any alterations, especially to open the packing of the goods, unless and to the extent to which it appears necessary in the public interest or in order to avert or minimize the loss.

§ 94.

Partial loss or damage.

(1) Where only some component parts or appertenances have been lost or damaged, the insurer is not liable but for the loss or damage done to the parts or appertenances aforesaid.

(2) The rule of subs. 1 does not apply, where experts have ascertained, in accordance with the rules of § 74, subss. 2 to 10, that, on account of the damage sustained through the perils insured against, the goods cannot be repaired or the costs of repair would exceed the insurable value of the thing insured; in such a case, the damage must be ascertained in accordance with the rules of § 93.

§ 95.

Alteration of means of transport.

(1) The insurer is not liable, if the goods have not been carried in the ship designated in the contract of insurance.

(2) The insurance continues if, after the risk has attached, the goods have been forwarded by land or in an other ship than the one designated in the contract of insurance on account of an accident by perils insured against; the person effecting the insurance, if circumstances permit him to do so, must ask the insurer for instructions as to the forwarding of the goods and must comply with his instructions, as far as he is able to do so. The same rule applies if, after the risk has attached, the means of conveying the goods have been altered or the ship has abandoned her voyage without the consent of the person effecting the insurance.

(3) In the case of subs. 2 the insurance covers the costs of transshipment, of temporarily warehousing the goods and the extra costs of forwarding them.

§ 96.

Sale of goods.

(1) Where, after the risk has attached and without the insurer being discharged from his liability, the ship has abandoned her

voyage or for any other reason the voyage has not been completed, the insurer is entitled to ask the person effecting the insurance to sell the goods with his assistance either off hand or by public auction, if the goods cannot be forwarded within a reasonable time or without incurring unreasonable expenses. If the goods must be sold at the request of the insurer, the sale must take place forthwith. The risk terminates with the goods being sold.

(2) Where the goods have been sold, the person effecting the insurance is entitled to claim the difference between the insured value and the proceeds of the sale, deduction being made of any sums he has received from third persons on account of the loss or damage sustained. This rule applies to the case where the goods had to be sold in the course of the voyage on account of an accident by perils insured against. Also in this case the risk terminates with the goods being sold.

(3) If by the rules and regulations of the sale the seller has to advance money in order to cover the expenses thereof, the insurer, provided he assented to the rules and regulations aforesaid, is answerable for the purchase-money being paid.

§ 97.

Floating policies.

(1) Where, before any insurable interest has attached, goods without any specification or certain classes of goods have been insured subject to being specified (being declared on the policy) after an insurable interest has attached, all goods or all goods of the class designated in the contract are held to be covered by the contract of insurance which the person effecting the insurance is bound, by the general rules of the law merchant, to cover by insurance either for his own or for another's account; goods, however, which the person effecting the insurance was bound to insure by special agreement, even if he is specially paid for on this account, are not held to be covered.

(2) The insurance covers also the goods mentioned in § 80, subs. 3; the insurer, however, is entitled to claim an additional premium.

(3) The insurer is bound to deliver to the person effecting the insurance a document signed by himself and embodying the contract of insurance (floating policy). A floating policy is not a policy within the meaning of the law or of these rules; § 15, however, of these rules applies with respect to the approval of the contents of the policy. Where the floating policy has been

lost or destroyed, the person effecting the insurance is entitled to ask the insurer to supply him with a copy of the policy; the costs of issueing a copy are to be born by the person effecting the insurance.

(4) The insurer, if asked to do so, shall deliver to the person effecting the insurance, every time he makes a declaration on the policy, a document signed by himself and embodying the declaration (special policy). This special policy is a policy within the meaning of the law and of these rules; § 15, however, of these rules does not apply to the special policy.

(5) The premium, including additional charges, is due not upon the conclusion of the contract, but at the date at which the risk attaches.

(6) The person effecting the insurance must declare the goods on the policy as soon as possible and, in particular, immediately after having received information as to the risk having attached; the declaration must contain the insurable value of the goods and the name of the ship, in which the goods are to be or are carried. The duty of the person effecting the insurance to comply with these rules does not cease even if he becomes aware, after the goods have already been safely delivered, of an insurance having been effected. The insurer is discharged from liability, if the declaration is not made within due course; it is, however, not necessary that the declaration should be communicated, it is sufficient that it has been sent off in due course, in order to maintain the liability of the insurer. The fact of the person effecting the insurance intentionally omitting to make a declaration or not making a declaration within due course or making a wrong declaration in respect of the goods or of their insurable value vitiates the insurance; the insurer, however, is entitled to claim the amount of premium due in case the contract had been duly fulfilled.

(7) Where it has been agreed that a ship shall carry only a certain quantity of goods the insurable value of which does not exceed a limited amount, goods, the insurable value of which exceed the aggregate amount, are not covered. This rule, however, does not apply where, without the person effecting the insurance being responsible for it, goods have been shipped at a port of transshipment thereby exceeding the amount agreed to.

(8) The insurer is discharged from liability, where goods have been carried in ships inferior to those designated in the contract, unless this has been done without the consent of the person

effecting the insurance. In this last case, the insurer is entitled to claim an additional premium.
(9) After an accident has happened which is covered by the insurance, the insurer is entitled to cancel the agreement by giving 4 weeks notice, even if the insurance has been taken for a fixed period of time. The insurer is debarred from cancelling the contract, if he does not do so immediately after having been informed of the accident and its consequences. These rules apply likewise to the person effecting the insurance, who is entitled to cancel the agreement in case the insurance has been taken for a longer period than one year and, after the first year has elapsed, an accident occurs by perils insured against.
(10) Where the insurance has been taken against risks of war only, each party is justified in cancelling the agreement by giving 3 days notice, even if the contract has been concluded for a fixed period of time. Where the insurance has been taken against other risks also, only that part of the contract can be cancelled which deals with the risks of war; in such a case, the premium is reduced by the amount calculated for covering the risks of war.

§ 98.

Open policies.

Where a policy has been taken out for a fixed period of time upon a fixed aggregate amount, of which the sums insured upon each declaration are to be written off (open policy), the insurer is entitled to claim a consideration within the limits defined in § 18 on account of the amount which has not been written off within the period of time fixed in the contract. Where a time limit has not been fixed in the contract, the insurance is considered as having been taken for 12 months. Where the open cover has been taken out for a period exceeding 12 months, the aggregate amount insured is being reduced at the end of every year by not less than the amount corresponding to the proportion of one year as compared with the full term of the open cover, the insurer being entitled to claim a consideration within the limits defined in § 18 in proportion to the sum not written off.

§ 99.

Other insurances in respect of goods.

Unless otherwise provided, the rules of section 2 apply also to other insurances covering any interest in or with respect to the goods, for instance to an insurance on prospective profits or commission or to a special insurance on freight paid independently of delivery.

SECTION 3.

Insurance on prospective Profits and Commission.

§ 100.

Valuation.

(1) Where, in an insurance on prospective profits, the parties have not agreed as to the value to be placed upon the subject insured, the sum insured is considered to be the value agreed upon.
(2) The insurer is entitled to have the valuation corrected, if the value agreed upon exceeds the amount of profits which, at the time the contract of insurance was entered into, merchants could possibly have expected to be forthcoming.

§ 101.

Goods and profits insured in one sum.

Where goods and profits have been insured in one sum, the tenth part of this sum is considered to be the insurable value of the profits insured. Where the parties have agreed as to the value to be placed upon the subject-matter insured, the parties are held to have agreed upon the tenth part of their valuation as being the value of the profits insured.

§ 102.

Floating policy.

The person effecting the insurance is entitled to declare upon a floating policy an amount of profits exceeding the amount stated in the contract. The insurer, however, is discharged from his liability for the amount exceeding, if the person effecting the insurance, at the time he made the declaration, knew or ought to have known that the voyage did not come to a successful end. In the case of the declaration being made for account of another or through an agent, the rules of §§ 22 and 57 apply.

§ 103.

Insurance on profits in case of loss or damage done to goods.

(1) There is a total loss of goods, where the goods do not reach their place of destination, whether it be on account of an actual total loss or for any other reason.

(2) Where the goods have been sold in the course of the voyage or, where the person effecting the insurance has been indemnified in virtue of §§ 611 and 612 German Commercial Code*) for goods totally or partially lost or disposed of for account of the shipowner in the course of the voyage, the amount, by which the proceeds of the sale or the sum paid in accordance with §§ 611 and 612 aforesaid exceed the insurable value of the goods, must be deducted from the sum insured.
(3) Where goods have been damaged, the person effecting the insurance is entitled to claim such a proportion of the sum insured, as the difference between the sound and damaged value bears to the sound value of the goods (§ 93, subs. 1).

§ 104.

Commission.

The rules of this section apply to the case of an insurance on commission to be earned upon the arrival of the goods at their place of destination. Where, however, the goods and the commission to be earned have been insured in one sum, 2 per cent. only of the sum insured is to be put to the account of the insurance on commission (vide § 101).

SECTION 4.

Insurance on Freight, Ship's Hire and Passage-money.

§ 105.

Insurance on freight. Measure of indemnity.

(1) With respect to the measure of indemnity in an insurance on freight, unless otherwise provided in this section, the rules of the

---

*) § 611. If a loss or damage has arisen, for which the owner is responsible, the cargo-owner is entitled, in case of total or partial loss, to claim a sum equal to the trading value, or if there is no trading value, a sum equal to the ordinary value which goods of the same kind and quality possess at the port of destination when discharge was begun; if no discharge took place, the value shall be fixed upon the time of the arrival of the ship; deduction shall be made of all savings effected through the loss, as customs, freight and other expenses.

If the ship does not reach the port of destination, the place shall be taken in substitution thereof at which the voyage comes to an end, or in case the ship be lost, the place to which the cargo has been brought into safety.

§ 612. For goods disposed of in the course of the voyage for account of the shipowner (§ 541 Commercial Code), the cargo-owner may claim the same amount as for goods lost; if the goods were sold for a greater amount, he may claim the latter.

first section relating to insurances on ship, of § 85, subs. 1, relating to cargo stowed on deck and the rules of § 86 relating to loss or damage through inherent vice or other like causes apply.

(2) For accidents occurring to the ship the insurer is not liable, unless and in so far as the contract of affreightment had already been concluded or, where goods have been shipped for account of the shipowner, these goods were on board when the accident occurred.

(3) Where goods have been shipped which are liable to melting in water, the amount of freight due, unless otherwise provided, is to be determined upon measure, weight or quantity delivered.

(4) In an insurance on time, the insurer on freight is not liable for any loss or damage occurring in the course of the voyage of the ship to her port of loading.

(5) Where time-freight has been insured, the insurer is not liable for any loss of freight caused through a delay in the commencement or in the prosecution of the voyage.

§ 106.

Insurance on freight. Duration of risk.

As regards the perils the ship is exposed to, the risk commences and terminates in accordance with the rules of the first section relating to insurances on ship. As regards the perils the goods are exposed to, the risk commences when the goods are taken on board; the termination of the risk is regulated by the rules of the second section relating to insurances on goods.

§ 107.

Insurable value of freight.

(1) The insurable value of the freight is the amount of freight stipulated in the contract of affreightment or, where no agreement has been made as to the amount of freight due or where goods have been shipped for account of the shipowner, the customary amount due at the place and time of shipment.

(2) In an insurance on net-freight. the insurable value consists of two—thirds of the gross-freight.

(3) Even if the parties have agreed as to the value to be placed upon the freight insured, the value agreed upon is not considered to be a valuation within the meaning of these rules, but merely a sum insured in an unvalued policy.

§ 108.
Ship's hire.

(1) Where ship's hire has been insured, the insurer is not liable for any loss caused through a delay in the commencement or in the prosecution of the voyage.
(2) The insurable value of the ship's hire is the amount of hire agreed upon in the contract letting the ship or, where no agreement has been made as to the amount of hire due, the customary amount.
(3) Even if the parties have agreed as to the value to be placed upon the hire insured, the value agreed upon is not considered to be a valuation within the meaning of these rules, but merely a sum insured in an unvalued policy.

§ 109.
Passage-money.

(1) Subject to the provisions in subs. 2 and 3 of this paragraph, where passage-money has been insured, the rules of §§ 105 to 107 relating to an insurance on freight apply.
(2) The commencement and the termination of the risk are regulated in accordance with the rules of the first section relating to insurances on ship.
(3) The insurer is also liable for loss or damage arising out of the fact that, by an accident through perils insured against, the person effecting the insurance is compelled to disembark the passengers at an intermediate port, to supply them at that port with board and lodging in pursuance to the rules of the law relating to the conveyance of passengers, to convey them to their place of destination by other means than those stipulated for in the contract and to indemnify them for loss or damage done to their luggage.

SECTION 5.
Disbursements.

§ 110.
Insurable value. Measure of damages. Subrogation.

(1) The insurable value of a claim secured by property exposed to marine risks is the amount of the claim plus legal or agreed interest; where a loan on bottomry or respondentia have been insured, the premium must be added to the amount of the loan. If the value of the property at risk was not worth so much at the time the risk attached, the insurable value is the value of the property at risk.

(2) The insurer is liable only in so far as the claim, on account of an accident by perils insured against, cannot be satisfied out of the property by which the claim has been secured.
(3) To the extent to which the loss has been paid by the insurer, he is subrogated to the rights and remedies of the person effecting the insurance in respect of his claim. The rules of §§ 45 and 46 apply.

§ 111.

Measure of indemnity.

Unless otherwise agreed to, disbursements are considered to have been secured by ship and freight, and loans on bottomry by ship, freight and cargo.

§ 112.

Duration of risk.

In case of bottomry or disbursements the risk attaches at the moment the person effecting the insurance has bound himself to advance the money. Where, in a case of disbursements, he himself has advanced the sum, the risk attaches at the time the money has been paid.

THIRD PART.

Special Clauses.

§ 113.

Free of particular average.

The insurer is not liable for damage nor, in an insurance on goods, for the fact that the damage done to the goods has resulted in their total loss or in the goods having been reduced to such a state as to be no longer capable of use under their original denomination. The insurer, however, is liable for damage done in order to save ship and cargo from perils insured against (§ 29).

§ 114.

Free of particular average, unless the ship be stranded.

(1) The insurer is not liable for any damage, unless the ship be stranded. The rules of § 113 apply; a damage, however, is considered to have been caused by stranding, if the facts of the case warrant such a conclusion.

(2) There is a case of stranding, where the ship has struck the ground and cannot be brought off but by taking extraordinary measures; cutting away of masts, jettison or the discharge of cargo. or waiting for the setting in of an unusually high tide are considered to be extraordinary measures; not so, however, the heaving of the ship on her anchors or backing the sails or the screw. There is a case of stranding also, where the ship has capsized, sunk, foundered, collided with other vessels, or has been fired upon or, where fire breaks out or an explosion occurs on board the ship.

(3) In an insurance on goods, there is a case of stranding, where the ship has struck or settled down upon the ground or has collided with other objects or has been damaged by ice or, where one of the accidents mentioned in the last sentence of subs. 2 has occurred. The insurer, however, is liable only in case the hull, through the stranding of the ship, has been damaged to such an extent that it can be inferred that also the goods have suffered from the accident. Where it is uncertain, whether or not a loss or damage by fire has been caused by spontaneous combustion of the goods, the person effecting the insurance must prove that selfcombustion was not the cause of the loss.

§ 115.

Free of breakage.

The insurer is not liable for damage caused by breaking. The rules of § 113 apply.

§ 116.

Free of breakage, unless the ship be stranded.

The insurer is not liable for damage caused by breaking, unless the ship be stranded. The rules of § 114 apply.

§ 117.

Free of average under a certain percentage.

The rules of § 34 apply, relating to the liability of the insurer in case the loss or damage sustained does not amount to 3 per cent. of the insurable value.

§ 118.

Free of the first (specified) percents.

The insurer is liable for any loss or damage sustained only to the extent, to which the loss or damage exceeds the percentage agreed upon in the contract of insurance.

§ 119.

Free of disturbances arising out of war ("Kriegsmolest").

The insurer is discharged from liability from the moment war begins to operate upon the course of the voyage, especially where the commencement or the prosecution of the voyage is impeded by men of war, privateers, mines or blockade or, where the voyage is postponed owing to war or where, owing to the same cause, the ship deviates from her course or the master is no longer free in the navigation of the ship.

§ 120.

"On safe arrival." "On safe course."

(1) Where an insurance has been taken on the safe arrival or on a safe course of the thing insured, the risk terminates when the ship has moored at anchor at her proper place at her port of destination.

(2) Where an insurance has been taken on the safe arrival of the ship, the insurer is only liable if, before the time mentioned in subs. 1, the ship has been totally lost or, where she has been abandoned in pursuance to the rules contained in §§ 72 and 73 or, where the ship has been sold by public auction in the cases mentioned in § 77.

(3) Where an insurance has been taken on the safe arrival of the goods, the insurer is liable only to the extent to which the goods do not arrive at their place of destination.

(4) The insurer is not liable for general average contributions or general average sacrifices nor for the expenses and costs mentioned in §§ 32 and 95, subs. 3.

§ 121.

War risks only.

(1) Only risks of war are covered by the insurance. The insurer, in particular, is liable for all loss or damage caused through measures taken on account of the war by a belligerent Power, whether the Power has been recognized as such or not, especially through the thing insured having been stopped, captured, taken, detained. requisitioned, put under restraint or having been damaged or lost through mines laid on account of the war or through other measures. A measure is considered to have been taken by a belligerent Power, if this Power joins in the war, raged by another Power, within six months after the measure has been taken.

(2) Subject to the foregoing rules and to the provisions contained in subss. 3 to 7 of this paragraph, the rules of § 35 apply.

(3) The insurer is not liable for expenses, not even for general average expenses, incurred on account of the ship, owing to the war, not commencing or not prosecuting her voyage, or calling at a port, or on account of the goods having been discharged, warehoused or conveyed by other means to their place of destination.

(4) The insurer is not discharged from his liability in the cases mentioned in subss. 2 and 3 of § 35; in the case mentioned in the last sentence of subs. 3 of § 35, however, the insurer is not liable for a subsequent damage or leakage, unless the ship be stranded.

(5) In an insurance on ship, the insurer is not liable for any loss or damage sustained on account of contraband goods being on board, unless the person effecting the insurance neither knew nor ought to have known that such goods were on board. In this last case, the insurer is not discharged from his liability: he is, however, entitled to claim an additional premium.

(6) There is a total loss in accordance with the rule of § 71, subs. 2, where a ship has been condemned as good prize. The rules of § 72, which apply to the cases mentioned in § 73, apply, in particular, to the case where a ship has been stopped, captured, taken, detained, requisitioned or has been put under restraint by a belligerent or other Power, whose measures, according to the rule in subs. 1, are considered to be measures taken by a belligerent Power; six months, however, must elapse, before a settlement as for a total loss can be asked for.

(7) An insurance on any kind of goods and merchandises does not include goods which are liable to capture and seizure as being contraband goods at the time the risk attaches. This is especially true in the case of a floating policy. The insurer, however, is liable for any loss or damage done to such goods, if the person effecting the insurance neither knew nor ought to have known that the goods were contraband goods; the insurer, in such a case, is entitled to claim an additional premium.

§ 122.

War risks included.

(1) The insurance covers also the risks of war.

(2) Where a concealment or a misrepresentation of material facts or an alteration of risk, which, under ordinary circumstances,

would have discharged the insurer from liability, has taken place solely with regard to or on account of the risks of war, the insurer remains liable for all other risks (§ 35); but, where the concealment, misrepresentation or alteration of risk bears reference to the other risks only, the insurer remains liable for the risks of war (§ 121), unless it can be inferred that he would not have taken the risks of war alone for the same consideration.

(3) In all other respects the rules of § 121 apply, subject, however, to the right of the insurer to claim an additional premium in accordance with the rule contained in § 35, subs. 4.

§ 123.

For total loss only.

The insurer is liable only for a total loss or where the thing insured is considered to be totally lost pursuant to the rule contained in § 72 or where it has been seized by authority or has been taken by pirates or, in an insurance on ship, also, where the ship has been sold by public auction in accordance with the rules contained in § 77. The insurer is not liable for general average contributions or general average sacrifices nor for the expenses or costs mentioned in § 32 and § 95, subs. 3; nor is he liable, in an insurance on goods, for the fact that the damage done to the goods has resulted in their total loss or in the goods having been reduced to such a state as to be no longer capable of use under their original denomination.

§ 124.

From warehouse to warehouse.

(1) The risk attaches, when the goods, at the place of shipment, are taken away from the place, where they have been kept, in order to be conveyed on the insured voyage.

(2) The risk terminates, when the goods, at the place of delivery, have been brought to the place, at which the consignee has ordered them to remain. As regards loss or damage sustained by fire, explosion, lightning or earthquake, the risk terminates not later than on the tenth day after the goods have been discharged; this rule applies also to the case where, by other reasons, loss or damage has occurred, unless the goods have been conveyed to the place, where they shall remain, without delay.

(3) These rules apply also in so far as the goods are conveyed over land or inland waters.

FOURTH PART.

Miscellaneous.

§ 125.

Sea- and inland-voyage combined.

Where an insurance has been taken for a voyage partly on sea partly over land or by inland waters, these rules apply to the whole transport including the conveyance by land or inland waters.

§ 126.

German Law to be applied.

All relations and differences arising out of the contract of insurance are governed by German Law. In so far, however, as the rules of the Statutes relating to insurances of transports are not binding upon, but may be altered by agreement of the parties concerned, the Statutory Law does not apply.

§ 127.

Jurisdiction.

All differences arising out of the contract of insurance are to be decided by the Court, within whose jurisdiction the insurer resides. Where, however, the insurer or his agent have a place of business outside the jurisdiction of the aforesaid Court, and a contract of insurance has been concluded by the agent of the insurer at this place of business, the aforementioned differences may also be settled before the Court, within whose jurisdiction the place of business is situated.

INDEX.

Numbers denote paragraphs and sections.

Numbers denote paragraphs and sections.

Numbers denote paragraphs and sections.

Numbers denote paragraphs and sections.

Numbers denote paragraphs and sections.

Numbers denote paragraphs and sections.

Bei Fragen zur Produktsicherheit wenden Sie sich bitte an:
If you have any questions regarding product safety,
please contact:

Walter de Gruyter GmbH
Genthiner Straße 13
10785 Berlin
productsafety@degruyterbrill.com